Studying Mathematics

Mary Catharine Hudspeth
Pennsylvania State University

and

Lewis R. Hirsch
Rutgers University

**KENDALL/HUNT
PUBLISHING COMPANY**
Dubuque, Iowa

Printed in the United States of America   **B**   402768   01

Contents

Preface

Many students feel intimidated by mathematics and want concrete suggestions on how to become better learners. This book is designed to meet these specific needs. The study techniques outlined here are based on the authors' many years of experience in working with both non-traditional and traditional students in courses as diverse as introductory arithmetic and algebra, and "math for the liberal arts." The students who have used the techniques suggested here have found them invaluable in helping them become more organized learners of mathematics. The book can be used independently by students, in a traditional class, or in math study skills workshops.

The authors wish to acknowledge their indebtedness to the many students with whom we have discussed these techniques and who have helped us refine them. In particular, we would like to thank Lura Stoedefalke and Susan Young for sharing their insights into learning mathematics. We are also indebted to Professor Frank Kocher of Baylor University; Professor Geoffrey Akst of the Borough of Manhattan Community College, CUNY; and Susan Schiller, Director of Learning Skills Center at the University of Pittsburgh, for their many helpful suggestions.

Introduction

You are probably reading this booklet for one of several reasons. Perhaps you lack experience with mathematics, yet know you have courses to take and wish to do well in them. Some of you feel intimidated because you have heard so much about the difficulty of mathematics. The thought of algebra or trigonometry, may, for instance, induce a mild (or not so mild!) state of panic. Others of you have that vague feeling that "you could do better" in mathematics than you have in the past, yet you do not know how to change your habits so that you can improve. Still others have had difficulty with mathematics and wish to find a way to "make mathematics go better." You may have spent a great deal of time on math and still have done poorly. You *all* have experienced frustration and, at times, may have felt defeated by mathematics, yet because you are looking at this booklet, you show that you are interested in strengthening yourself in math.

People like yourself who have had these concerns about learning math have been helped by following our suggestions for study. They have become better organized and they have gained confidence in mathematics. They were able to accomplish this because they **WORKED** and **WORKED HARD** at learning math. Unfortunately, reading this booklet without any commitment on your part will not bring change, but using it to help yourself make better use of your time **WILL** help. We do not claim you all will be able to become mathematicians, but all of you can become better students of mathematics, and the vast majority of you *can* do well in freshman-level or non-major courses for which you have the background.

Math Study Skills Evaluation

This checklist will help you measure your study habits in mathematics. Respond to each item as honestly as possible.

	Seldom	*Sometimes*	*Usually*
1. Before I begin to do my math homework, I read over my notes or mathematics text.	_____	_____	_____
2. When I miss a class, I get the notes from someone else in class.	_____	_____	_____
3. Whenever I run into a mathematics problem I have trouble with, I stop studying.	_____	_____	_____
4. I make up the assignments when I miss a math class.	_____	_____	_____
5. I prefer not to ask questions in math class.	_____	_____	_____
6. I don't look over the homework or quiz problems I get wrong.	_____	_____	_____
7. When I get confused in math class, I stop taking notes and think about something else.	_____	_____	_____
8. If I have a lot of difficulty with a topic in math, I go to see the instructor or tutor for help.	_____	_____	_____
9. I do most of my studying for a math test the night before the test.	_____	_____	_____
10. I only look at the examples in my math book, I don't read the book.	_____	_____	_____

	Seldom	*Sometimes*	*Usually*
11. I study in a quiet, well-lit place away from distractions.	_____	_____	_____
12. When I see a word I don't understand in my math text, I look it up.	_____	_____	_____
13. Given a choice, I'd rather sit toward the back of the room in math.	_____	_____	_____
14. I review the material covered in the course even if the teacher does not assign the review.	_____	_____	_____
15. On the average, I study between two and three hours for each hour I spend in math class.	_____	_____	_____
16. I take notes in math class.	_____	_____	_____
17. Working assignments is the only math studying I have time for.	_____	_____	_____
18. If I have time, I try to check my answers on exams.	_____	_____	_____
19. On a math test I start with the first problem and work straight through the test.	_____	_____	_____
20. I do most of the math in my head and don't have to write down many steps.	_____	_____	_____

See page xi for Scoring Guide for the Math Study Skills Evaluation.

How do you use this book?

Most of you will choose to read the whole book in order, but some of you may have specific problems that you wish to work on first. In order to guide you to the appropriate sections, we have listed the most common problems and the pages on which they are discussed.

Scoring Guide for the Math Study Skills Evaluation

You can score yourself by giving yourself points as follows:

Items 1, 2, 4, 8, 11, 12, 14, 15, 16, 18	_____ 5 points for each "Usually" _____ 3 points for each "Sometimes" _____ 1 point for each "Seldom"
Items 3, 5, 6, 7, 9, 10, 13, 17, 19, 20	_____ 1 point for each "Usually" _____ 3 points for each "Sometimes" _____ 5 points for each "Seldom"

TOTAL _____

If your score is below 70, you can markedly improve your math study habits.

If your score is between 70 and 85 you have many good habits, but you can improve your skills.

If your score is above 85 you have excellent math study skills.

CHAPTER 1
A Perspective on Mathematics

What makes mathematics different?

> "But I do okay in *other* subjects . . ."

Before we discuss how to study mathematics, we need to understand what mathematics "is"—that is, what makes it different from English, history, or biology, for instance. Many of you do well in these other subjects, but find that mathematics does not come so easily. There are several reasons for this. While all of the subjects require work, study, and thought, much of the material in the other subjects is related to what you encounter in your daily life: as you watch TV, listen to the radio, or read a magazine or newspaper you may learn about biology, government, or psychology. You are not likely to hear about adding fractions or solving a quadratic equation. In many nontechnical courses you may use this general information in class discussion or on examinations. In technical courses like mathematics, however, you must not only learn the necessary facts presented, but you must learn the related procedures. For instance, knowing the *fact* that fractions with unlike denominators can be added will not earn you a passing grade in a mathematics class— you must also know the necessary *steps* needed to actually *add* the fractions.

An even sharper contrast between mathematics courses and other subjects occurs when we consider the use of personal opinions. In English you may write a paper presenting your thoughts about the appropriate legal age to buy alcohol. In mathematics you are not asked for your opinion about the appropriateness of adding fractions; rather, you must learn the techniques that are appropriate to use.

In a history course you may miss a week or more of class, yet be able to understand much of the discussion when you return because the class has moved on to new topics. In most mathematics classes, however, if you miss a week, you may not be able to understand the discussion when you return because the discussion presumes that you have learned all of the material you have missed. Mathematics is sequential—you must understand the earlier material before you can learn the new. For instance, you must know how to add numeric fractions (i.e., $3/14 + 1/5$) before you can learn how to add algebraic fractions (i.e., $1/x + 3/2x$).

Learning mathematics is much like learning a language or learning to play a sport or a musical instrument. Mathematics *is* a language: you must learn the vocabulary—the terms and definitions used. Words that have one meaning in ordinary conversation, like "factor," "power," "solution," and "term," have a much more specific and precise meaning in mathematics. Other words are not even a part of daily language—words like "quadratic," "numerator," and "polynomial." You also must learn the "grammar" of mathematics—the way in which the symbols can be used to make meaningful statements. For example, "$5 + 7 = 12$" has meaning, but "$2(\times \div 3) - = \sqrt{+}\ 9$" does not. Of course, you may be able to understand a language without being able to speak it. But, the only way to become fluent is to practice speaking until you can talk without having to stop to think of the word or of the correct structure to use. To *know* mathematics is to be able to "speak mathematics" fluently, which requires extensive practice.

The second example may be meaningful to more of you. In order to become a good musician or a sports competitor, you must practice regularly and keep challenging yourself with more difficult tasks. As a young child you would practice half an hour every day; as you grew older and your skills increased, your practice time lengthened until many of you practiced several hours *every* day. When you resumed practice after skipping for several days (or more), your skills were off; you had to start at a lower level. Some of you may have taken up a new sport as an adult. You can recall how slowly you had to begin and how long it took before you became proficient, no matter whether it was jogging, swimming, or racquetball. In order to develop musical talent or sports ability, you must discipline yourself to daily practice over an extended period of time. You must do the same if you expect to develop your mathematical ability. You need sufficient practice so that the mathematics becomes "automatic," so that you do not need to stop and question the next step, you *know* it.

Why are college courses harder?

All of the preceding comments about mathematics are true no matter what the level—junior high, high school, or college. However, there *are* differences between a college course and one in high school. In general, there are two major changes. Because the college course meets for fewer hours a week, each class meeting covers more material and, second, the students are given more responsibility for learning it. In most college courses you will *not* have the opportunity to take a retest when you do poorly on an exam. Usually, a college student is expected to study *three hours* for *every* hour in class. Because not every high school student goes to college, competition in a college class is frequently greater. College instructors also have less reluctance about giving grades of D or F. Thus, you can see that people who "got by" math in high school frequently feel very frustrated in a college course. One simply **MUST** be better organized in order to succeed.

That is of course, where your desire to learn can make a difference. If you did not learn the necessary material in junior high or high school, you know you want to learn it now so that you can succeed in college. With our guidance, you can develop better study habits and begin to develop the characteristics of a good student of mathematics.

The Weak Student

1. Does not plan ahead.

2. Gives up when does not understand.

3. Does not attempt to find any relationship among similar problems.

4. Stops studying when homework is done.

5. Attempts to memorize all rules without understanding *why* the rules work.

The Good Student

1. Is organized.

2. Knows appropriate questions to ask to clarify the material.

3. Knows how to focus on key steps and procedures.

4. Continues to study until the appropriate procedures and techniques become almost automatic.

5. Thinks through the problems rather than trying to rely solely on memorized procedures.

How do you choose an appropriate course?

You should strive for these goals, but in order for them to make a difference, you must be enrolled in a course that is appropriate to your skills and background. Far too many college students enroll in mathematics courses which assume greater knowledge than the students possess. The catalog may say "prerequisite one year of algebra" or "one-half unit of trigonometry," but high school courses are not uniform. Just because you took a course in which "trigonometry" was discussed does not mean that the course covered the necessary topics in sufficient detail to give you adequate command of the material. *Most schools have placement tests in mathematics to guide the students to appropriate courses. Use these test results to guide you.* If you are in doubt about your background to handle a particular course, by all means consult with the mathematics department, your academic advisor, or the placement center (or all three!). Students who take the initiative usually find people willing to answer their questions and to suggest appropriate math courses.

As you try to determine which math course is appropriate, you may find that your skills are not adequate for the required courses. You may need to take a preparatory course or two. Your school may require you to take some remedial work. Some of you resist "wasting your time" in a course that "does not count." For most of you, however, *not* taking the lower level course for background is a *bigger* waste of time: you enroll in the higher level course and then either drop it or fail it. Not only have you *not* passed *that* course that semester, you have *not* improved your skills so that you can take it with confidence the next semester. In addition, you have reinforced the pattern of not doing well in mathematics. The people who are more cautious and take the necessary preparatory courses earn better grades and have more confidence with mathematics.

After you enroll in a course, definitely discuss your preparation with the instructor if you have any doubts. If you find yourself in a math course for which you do not have adequate background, **DROP IT.** Each college has different regulations about dropping courses, so be sure to check the rules carefully. However, it is better to drop a course for which you are unprepared than to try to "stick it out" and undergo great frustration and reinforce your feeling that "you can't do math." If you do find yourself dropping a course, *be sure* to consult with your academic advisor and with your math instructor or the department to determine a more appropriate course.

What is the instructor's responsibility, and what is yours?

Most mathematics instructors are very willing to help their students. Most announce office hours when they are available for individual help. If you are not free during the announced hours, you can talk to the instructor after class or at another time convenient for both of you. Do not, however, expect the instructor to repeat the lectures or to do your studying for you. You should have thought about, and perhaps written down, your specific questions. There is nothing more frustrating to an instructor than to have a student come in for help with the statement, "I don't understand any of this stuff." The instructor has absolutely no idea what the student means by "this stuff"—it could mean all of Chapter 4, the last class, or a specific type of problem. The instructor is also likely to conclude that the student has not really tried to study the material to determine what he/she does and does not know.

In one of your first class meetings you should exchange phone numbers with a fellow classmate, someone you can call for the assignment if you have to miss class. The two of you might agree to pick up extra hand-outs for the other if one of you is out. If you know in advance that you must miss class, you should contact your instructor to learn the assignment and (if necessary) to make arrangements to complete any work missed.

When requesting help before a test, try to come in several days before the test itself. You are much more likely to get the individual attention you need by avoiding the last minute crush of students asking for help.

Study Questions for Chapter 1

1. Why is math different from other subjects?
2. In general, how often should you practice mathematics?
3. List several characteristics of a good student in mathematics.
4. How can you get the assignment when you have been absent?
5. Who can you see to discuss your placement in math?
6. When should you go to the instructor for help?
7. What should you do before going to the instructor for help?

Questions for the Student

1. Can you find a fellow student to study with?
2. What is the first change you think you should make in *your* study habits?
3. Do you know when your instructor or tutor is available to help you?
4. Do you know where you would go at your school to discuss your math placement?
5. Do you know your school's rules about dropping a course?

CHAPTER 2
Taking Notes in Class

"I follow everything that goes on in class, why do I need notes?"

The two main sources for information about mathematics are the textbook and the instructor in class. The text is a "permanent" record to which you can refer any time. While it is important to learn how to use the text, it is most important that you have an organized way of recording and studying the class activities if you are not in a self-paced class (each student working independently). For most classes, there is no way you can "replay" a class so that you can "take another look" at what happened in the same way you can "take another look at what the book says in Chapter 3." You can lay a good foundation for success in mathematics by being organized and prepared to take good notes for *each* class.

How can you be organized to make the best use of class?

The first thing to do is to make sure you have a specific notebook or section of a notebook to use for your mathematics course. A loose-leaf notebook is good because you can insert homework, class handouts, and tests in the appropriate spot. A notebook, however, does you no good unless you take it with you to **EVERY** class! Students who attempt to jot down problems on scraps of paper are not creating any permanent record to which they can refer later. Be sure also to have an adequate supply of sharpened pencils or pens with you. Some students also find it helpful to bring the text so that they may refer to it during class. Some instructors may require you to bring your text to class.

Just as it is important to have the necessary supplies for class, it is equally important to have **YOURSELF** ready for class. You need to be there on time for every class, reasonably rested, and prepared to concentrate on the lesson. *Do not miss class* unless absolutely necessary. This will save you study time outside of class and will improve your comprehension of the material.

If you can choose your seat, pick one in the front half of the room. On the average, the better students sit toward the front, and the students who do less well sit in the back. If you sit in the back part of the class room you are less likely to be able to hear the instructor and to read the work on the board. You also are more likely to be distracted from the material.

Many students find it helpful to read ahead in the text so that they know what topics will be presented in class. This way they can see what terms and/ or procedures will be new to them and what is more familiar.

Should you ask questions in class?

Once you have found an appropriate course, remember it is *your* course and that you are paying for it! The instructor is there to help you get the most from your investment. Do not be afraid to ask questions because you might look foolish. In general, if *you* have a question, then it is safe to assume at least three other people in class have the same question but are equally reluctant to ask! Ask it for them! Rest assured that the instructor will let you know when you are asking "too many" questions. If you are hesitant to ask in class, jot the question down then be *sure* to see the instructor after class or during his/her office hours. Those hours are to help YOU. Too many students somehow think it is a sign of defeat to go in to the instructor when, in reality, it is a sign of maturity.

What is the purpose of note-taking?

Let us examine the purpose your notes will serve in order to learn what form the notes should take. Most people need to develop two kinds of skills in mathematics:

1. The specific, detailed procedures needed to work a particular kind of problem.
2. The general principles that determine *what kind* of problem you have and *what procedures* are appropriate.

Thus, there are two different levels on which you need to focus in class. The first level is on the specific details of *which* steps to take *where* on a particular type of problem. The second level is on the *general rules* exhibited by the particular example. *In order to be successful in mathematics you must have command at both levels.*

> "I get lost in class and just give up."

Most people need time to think about new ideas and procedures before they understand them fully. Do not expect to follow completely every step that occurs in class. You need to record the "class proceedings" in order to be able to go back and refine your understanding so that you *can* follow all of the steps. *It is especially important to take detailed notes when you feel completely lost.* If you can simply record the examples and the key ideas even though you may not understand them at the time, you can then go back over the notes more slowly and try to understand them. If you do not take notes, however, you are losing one important opportunity to understand the material, and you are making your study just that much more difficult.

How can you organize your notes to record *both* types of information?

Obviously there are many different ways of organizing the way you record information from class. We will present several methods that successful students use to take notes.

Method I: Two-column system

In the two-column system of note-taking, the examples given in class are recorded in one column and the discussion of rules and procedures are recorded in the adjacent column. (See Figure 1.) You may choose to make each column one-half a page, or simply use the left page for discussion and the right page for examples. In general, you want to be generous with space when you take notes so that you can go back and insert steps and/or comments as you review. Figure 2 also illustrates notes taken using the two-column system. In both these examples, the key is to have the general rule opposite the specific step illustrating the rule. In Figure 1 the student found it helpful to draw lines directly connecting the term with the factor that it represents in the example. Also note the star that the student added to emphasize the point that rate and time must be expressed in the same time units. In Figure 2 a key concept has been starred. The comments written in the margin were added when the student went back to review the notes. Later we will discuss the use of the margin and what to write in it.

Method II: Integrated system

You may wish to have the examples and the discussion of the problems written one after the other, without splitting your page. In Figure 3, the student has chosen to copy the examples down and then "box in" the key rules that have been developed from the examples. The graph has been labeled with each of the new terms introduced in the class. The topic of the discussion is written at the top of the page along with the date of the class.

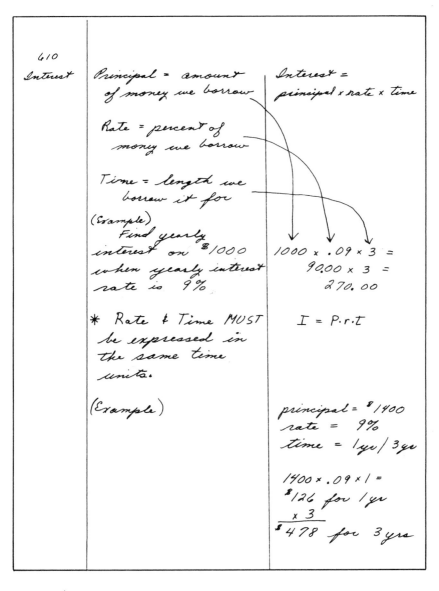

610
Interest

Principal = amount of money we borrow

Rate = percent of money we borrow

Time = length we borrow it for

(Example)
Find yearly interest on $1000 when yearly interest rate is 9%

* Rate & Time MUST be expressed in the same time units.

(Example)

Interest = principal × rate × time

$1000 \times .09 \times 3 =$
$9000 \times 3 =$
270.00

$I = P \cdot r \cdot t$

principal = $1400
rate = 9%
time = 1 yr / 3 yr

$1400 \times .09 \times 1 =$
$126 for 1 yr
× 3
$478 for 3 yrs

Figure 1

9

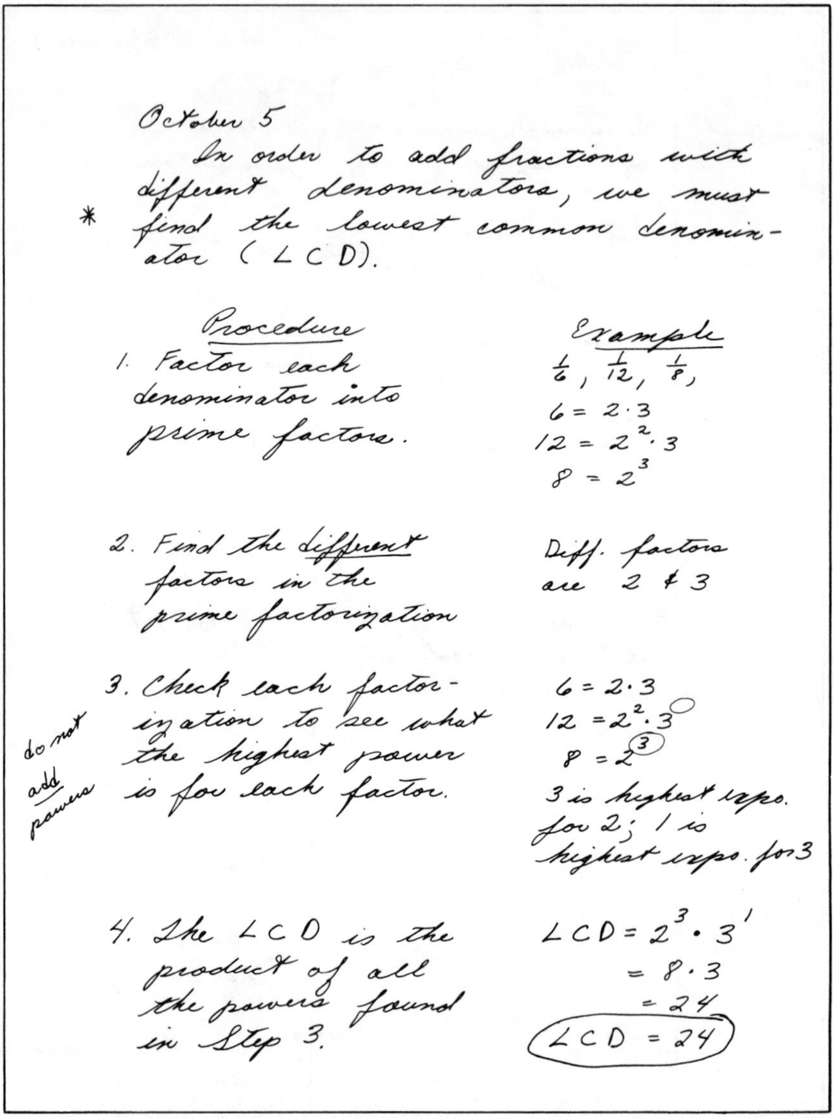

October 5

*   In order to add fractions with different denominators, we must find the lowest common denominator (LCD).

Procedure	Example
1. Factor each denominator into prime factors.	$\frac{1}{6}, \frac{1}{12}, \frac{1}{8},$ $6 = 2 \cdot 3$ $12 = 2^2 \cdot 3$ $8 = 2^3$
2. Find the different factors in the prime factorization	Diff. factors are 2 & 3
3. Check each factorization to see what the highest power is for each factor. *do not add powers*	$6 = 2 \cdot 3$ $12 = 2^2 \cdot 3$ $8 = 2^3$ 3 is highest expo. for 2; 1 is highest expo. for 3
4. The LCD is the product of all the powers found in Step 3.	$LCD = 2^3 \cdot 3^1$ $= 8 \cdot 3$ $= 24$ $LCD = 24$

Figure 2

10

Graphs

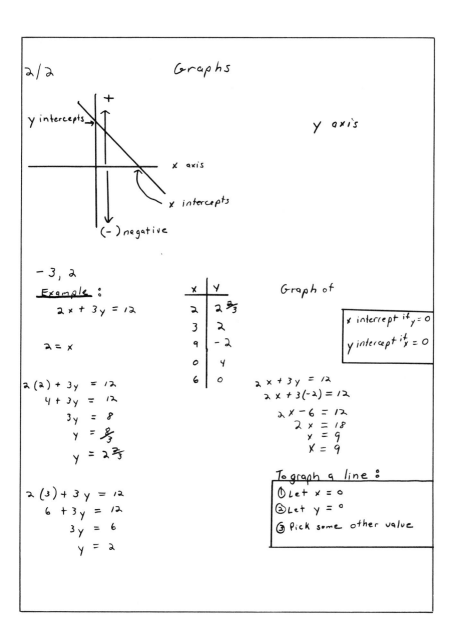

y intercepts→

+

y axis

x axis

x intercepts

(-) negative

− 3, 2

Example :

$2x + 3y = 12$

$2 = x$

x	y
2	$2\frac{2}{3}$
3	2
9	-2
0	4
6	0

Graph of

x intercept if $y = 0$

y intercept if $x = 0$

$2(2) + 3y = 12$

$4 + 3y = 12$

$3y = 8$

$y = \frac{8}{3}$

$y = 2\frac{2}{3}$

$2x + 3y = 12$

$2x + 3(-2) = 12$

$2x - 6 = 12$

$2x = 18$

$x = 9$

$X = 9$

$2(3) + 3y = 12$

$6 + 3y = 12$

$3y = 6$

$y = 2$

To graph a line :

① Let $x = 0$

② Let $y = 0$

③ Pick some other value

Figure 3

The notes in Figure 4 contain many abbreviations and symbols so the student could write the notes more quickly: "Relationships" is abbreviated to "Relships;" "Right triangle" is written "rt. △." Be sure, however, that your abbreviations contain enough information so that you can come back to your notes later and be able to understand them.

No matter which format you choose, your goal is to record the proceedings of the class in a manner that is clear and well-organized. One way to judge your note-taking is to ask yourself whether someone else in the class would be able to read and understand your notes. Some students actually compare notebooks in order to learn how other students organize the material.

Once you have a format chosen, WHAT do you write down?

Because you are working to increase your fluency in the "language" of mathematics, you need to record any definitions and new symbols that the instructor introduces. As all of the sample notes have shown, you also need to write down steps in the problems worked and record the rules for those steps. However, **YOU CANNOT WRITE DOWN EVERY WORD** that is said in class. You must choose the most important ideas to record. You may wish to use abbreviations, but make sure you know what your abbreviations mean! Better students record the date at the beginning of each class so that they will know the relationship between the notes and the classes. Obviously if you miss class, you should copy someone else's notes for that day.

> "I don't know what to write—class goes so fast."

In general, you will wish to record:

1. Any material written on the chalkboard. Especially include any examples as they are worked.
2. The new terms, symbols, definitions, and techniques presented.
3. All of the material that the instructor "high-lights" for you.

The only way instructors can "write in italics" or "write in capital letters" is by using verbal cues. They may:

1. Repeat information.
2. Summarize a section or chapter.
3. Use key phrases such as "Don't do this," "This is a common error," "This is important," "This is a key concept," "This distinction is important."

You should be *sure* to record all information that follows such cues.

As you take notes, you should try to follow the steps taken in examples. Good students try to anticipate what the next step will be. If you think you

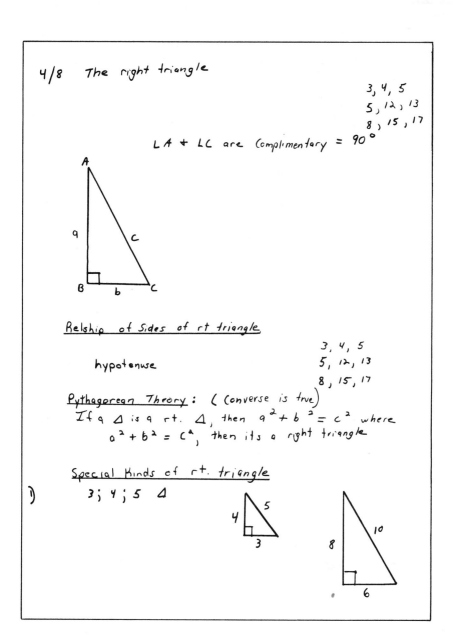

4/8   The right triangle

3, 4, 5
5, 12, 13
8, 15, 17

∠A + ∠C are Complimentary = 90°

Relship of Sides of rt triangle

hypotenuse

3, 4, 5
5, 12, 13
8, 15, 17

Pythagorean Theory : (Converse is true)
If a △ is a rt. △, then $a^2 + b^2 = c^2$ where
$a^2 + b^2 = c^2$, then its a right triangle

Special Kinds of rt. triangle
1)   3; 4; 5 △

Figure 4

13

are familiar with the material, you should try to work ahead of the instructor to verify that your steps and procedures are correct. If they are not, see if you can understand why they aren't. If you still have trouble, ask in class or make a note in the margin to ask after class.

Be sure to write down "too much" rather than "too little." Problems seem easy when someone else is working them. While you may *understand* the problems, you must have *command* of them. You must not only be able to *follow* the steps in a problem, you must be able to *provide* the steps yourself. You have command of the material when you can both recognize the type of problem and determine the steps necessary to solve it. Only by copying examples so that you may study them later can you be sure that you will gain *command* of the material.

Some people have great difficulty in taking adequate notes. This may be due to a hearing impairment, a weak background in the material, or an instructor who presents the material very rapidly or disorganizedly. In such cases some students use a tape recorder to supplement their notes. Because a tape recorder cannot copy the material from the chalkboard, *you* must copy it. If you choose to use a tape recorder, you may wish to make marks (comments) on your notes to indicate which comments were made with which example. In this way, you can better relate the two.

If the text is reasonably readable and if the instructor is following it fairly closely, try simply highlighting or underlining in your text as a form of note taking. This would save time and allow you to think more about what is being said.

Study Questions for Chapter 2

1. Give two main sources of information about mathematics.
2. In general, what material should be recorded in your notes?
3. How many questions should you ask in a math class?
4. What are some ways instructors let you know that something should be recorded in your notes?
5. What two types of skills should your notes help you develop?
6. When is it most important for you to take notes?
7. How often should you go over your notes?

Questions for the Student

1. Are you on time for class?
2. What are some ways *your* instructor lets you know that something should be recorded in your notes?
3. Is your notebook for math organized?
4. From your seat in class can you hear and see well? If not, how can you get a better seat?

5. Exchange last week's notes with a fellow classmate and see if you can understand what he or she wrote. Could you follow the notes? How do these notes compare with yours? Comment on the notes and ask him or her to do the same with yours.
6. Should you bring your text to every class?
7. How often do you go over your notes?

CHAPTER 3
Studying Your Notes and the Text, Then Working Assignments

Before we discuss *how* to study, we need to talk about *where* to study. Your place of study must be conducive to concentration and serious thought. Some of the factors that provide such an environment are:

1. A well-lighted table or desk with a comfortable chair (but not TOO comfortable!).
2. Quiet—**NO** television!! A place where roommates, friends, children, pets, etc. will not distract you.
3. An area (if possible) in which you *only* study, so that you will be less tempted to let your mind wander.
4. A good supply of paper and sharp pencils.

Some people study at a dining room table; others go to a library or a quiet lounge in a dorm; others have no difficulty studying in their room. No matter what your location, be sure you make it as conducive to concentrated work as possible.

As you study, be sure to give yourself a break of 5–10 minutes every 45 minutes or so. When you find you are losing your ability to concentrate, then switch subjects or put away your books! Some people prefer to get up early and study math for several hours while their mind is still fresh. The important point is that spending time with a math book propped up in front of you does you no good if you are not alert and able to concentrate. While it is crucial that you study mathematics every day, pick the time of day that will allow *you* to concentrate best.

Find someone in your class with whom you can study. Of course you will not always study together, but most students find it very helpful to have a "buddy" to talk to about mathematics. A "buddy" also is a person with whom you can exchange notes and assignments if one of you misses class. Many people find studying together for examinations particularly helpful.

Now that you have notes, how do you use them?

Review your notes after each class before you work on the assignment. Some people prefer to read the text prior to reviewing the notes. Here you must make a decision based on the particular class. In some cases, you will find it easier to read the text first and then use your notes to supplement the information from the book. In other cases, reading your notes first will allow you to understand the book more easily. In any case, you should use **BOTH** your notes and the text to help you understand the material **BEFORE** you try the assigned problems.

The first thing to do when reviewing your notes is to simply read them to find and mark definitions, new symbols or terms, and key ideas. Some people like to use a high-lighter pen (magic marker) when marking their notes. *Now is the time to use the margin!* In it you want to construct for yourself a basic outline of the material by recording key words there. You will want a key word for each item that you marked in your notes. That means you want a key word for each definition, new symbol or term, new procedure, and each different type of problem. Later, as you review for quizzes and examinations, you can cover the body of your notes and drill yourself on the key words you have recorded. You can then uncover your notes to see if you are correct.

You should also make out a study card now for the key words. (See Chapter 4 for the details on study cards.) For the notes shown in Figure 1, the student made out the note card given in Figure 5.

Next, reread the notes to work through the examples. As you work them, fill in any steps that were omitted. Make sure you understand not only all the steps involved in solving the examples given in class, but also the reason why those steps were taken. That is, ask yourself why "that" step was taken rather than another one. After you believe you understand the notes, try to work through the examples without the notes, then check your steps and the answer (if there is one given). Think about the strategies that you used and why they did (or did not) work. If you have questions about procedures in general or about specific steps, note your questions so that you can ask a fellow student or the instructor.

Third, skim the notes both to reread the definitions and to determine what additional key words you should add to the margin (recall column). Try to summarize in your own words the major points of the notes.

But what about the text?

Many students are tempted to begin work on the assigned problems without first reading the text. As we stated above, you must determine whether you learn better by reading the text before you review your notes or whether reviewing your notes before you read the text is more helpful. No matter **WHEN** you read the text, you should follow much the same procedure you use to review your notes.

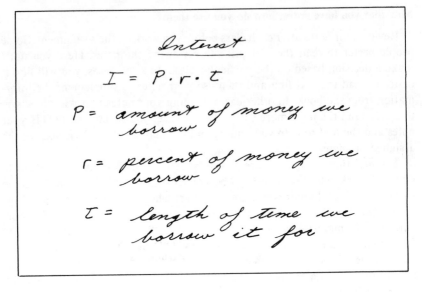

$$I = P \cdot r \cdot t$$

$P =$ amount of money we borrow

$r =$ percent of money we borrow

$t =$ length of time we borrow it for

Figure 5

First, skim the appropriate pages to mark definitions, new symbols or terms, and key ideas. Pay particular attention to the material that was **NOT** covered in your notes. Make note cards for any new material.

It is important to realize that your instructor may present the material in a very different way from the text. You should pay attention to any marked differences between the two presentations. Sometimes teachers deviate from the book for very specific reasons:

1. The teacher believes he or she has a better, easier, or more efficient way to approach the material.
2. The teacher wishes to give a broader perspective than does the book.
3. The instructor wishes to simplify the text's presentation.

Be sure you understand what was discussed in class because teachers tend to test on what *they* have said! If the two presentations seem to be contradictory, then be sure to talk to your instructor.

> "I can't *read* a *math* book!"

You are now ready to read the text thoroughly. It is important to realize that you cannot read a mathematics text as you do a novel. For much fiction, when you do not understand, you can simply keep reading; the relationships of characters become clear as you get farther into the book. That is exactly the opposite of the technique to use when reading mathematics. As soon as you encounter a word, term, or expression you do not understand, **STOP**

READING. Ask yourself what the author might mean. Is the term defined earlier in the book? If you still do not understand after thinking about the difficulty for several minutes, **WRITE DOWN YOUR QUESTION** or mark the book so you can ask. Only then should you continue on with your reading. This will be a SLOW process. Most people read novels at the rate of 20 to 40 or more pages an hour. You may read **TWO OR THREE PAGES OF MATHEMATICS** in an hour if the material is unfamiliar and difficult.

Try to work the examples and fill in steps that have been omitted. Focus on the reasons why each step was taken. If you do not understand, be sure to note your questions so that you may ask. After you have completely read the appropriate pages of the text, go back and try to work completely the examples **WITHOUT** looking at any of the solutions. When you have finished, check your answer and/or your steps. Try to reflect on the steps you took and why they did (or did not) work. Be sure to compare the different types of examples and the directions for each. If the directions are different, focus on the solutions that are appropriate for each set of instructions. Again, be sure to write down any questions that you may have.

Working the assignment

You may not realize it, but when you studied your notes and the text, you were preparing yourself to do your assignment. Those activities are an important part of your "homework"! If you skip that portion, you have *decreased* your efficiency in working the problems and you have *increased* the possibility that you will not completely master the material. So you see, taking time to study your notes and read the text saves you time and frustration in the long run.

Just as you should review your notes after every class meeting, you need to work problems each day. Mathematics is not a "spectator sport"! The only way you learn mathematics is the same way you learn bike riding or swimming: you must practice, and practice every day if you expect to master it. Whenever you fall behind, you have not only *not* increased your mathematical skills, but you have made the next class less understandable for yourself. For better or for worse, mathematics is a cumulative subject. What you learn in today's lesson you will need in order to understand tomorrow's class. Therefore, whenever possible **NEVER GET BEHIND** in mathematics!

> "But *I* understand the steps I took."

As you work problems, you are doing more than learning a procedure or getting an answer; you are establishing patterns of "communicating mathematics." Just as you practice writing sentences, paragraphs, and essays in an English class in order to communicate better in written language, so, too, you practice writing mathematics in order to communicate better in written mathematical language. Your goal is to communicate the steps taken and procedures used in a neat, clear, concise, organized way so that someone else can easily follow your work. If you establish good habits while doing your homework, you then make taking quizzes and tests less awesome. You will be more confident because you know how to present your work in an acceptable form.

Figure 6 illustrates the way one student organized her homework. The work was done in pencil so that erasures could be made and the paper would still look neat. She recorded the page numbers and problem number for easy reference. Each problem is given ample space, and the answer for each is circled. Because each step is clearly written, one below the other, we have no difficulty in following her work.

As you begin a set of exercises, first **READ** the directions! It is crucial that you familiarize yourself with the directions so that you understand them and so that you will recognize similar problems on quizzes and tests. You should be familiar with your text. Does it give the answers to any of the problems? Many texts give answers to the odd-numbered problems in the back of the book. Some of the less advanced texts give even more answers. You should try some of the problems (similar to the ones you have been assigned) for which the book gives answers. If you do not know how to begin, refer to the examples in your notes and/or the text. As you gain confidence, work a few more problems at a time before you check your answers. Next, work the assigned problems and/or the ones without answers given. Be sure to mark the problems that give you difficulty. At a later time take a second look at these problems; frequently you will discover that you *can* work them. If you find that you still do not understand the problems after restudying your notes and the text, be sure to seek help from a fellow student (your "buddy"), a tutor, or the instructor. It is crucial that you ask questions to clarify your understanding.

Don't stop studying just yet!

> "I work my homework, but it just doesn't 'stick'!"

You may wish to close your books and breathe a sigh of relief as you finish the last problem, but **DON'T STOP NOW!** You need to take a few minutes to reflect on what you have just done in order to consolidate your learning. First, if you have not yet done so, you should make note cards for any particularly difficult concepts.

Ch. 6 Sec. 4 pg. 284 (3,7,11,23)

3) $\dfrac{8x-5}{x^2-x-6} = \dfrac{8x-5}{(x-3)(x+2)} = \dfrac{A}{(x-3)} + \dfrac{B}{(x+2)} = \boxed{\dfrac{\frac{19}{5}}{(x-3)} + \dfrac{\frac{21}{5}}{(x+2)}}$

$8x-5 = Ax + 2A + Bx - 3B$
$\qquad = (A+B)x + (2A-3B)$
$8 = A+B \rightarrow 24 = 3A+3B$
$-5 = 2A-3B \rightarrow \underline{-5 = 2A - 3B}$
$\qquad\qquad\qquad \dfrac{19}{1} = \dfrac{5}{1}A$
$\qquad\qquad\qquad \dfrac{19}{5} = A$

$\dfrac{40}{5} = \dfrac{19}{5} + B$
$\underline{-\dfrac{19}{5} = \dfrac{-19}{5}}$
$\dfrac{21}{5} = B$

$\dfrac{40}{5} = \dfrac{19}{5} + \dfrac{21}{5}$
$\dfrac{40}{5} = \dfrac{40}{5}$ OK

7) $\dfrac{x^2-3x+4}{(x-1)(x+1)(x+2)} = \dfrac{A}{(x-1)} + \dfrac{B}{(x+1)} + \dfrac{C}{(x+2)} = \boxed{\dfrac{\frac{1}{3}}{(x-1)} + \dfrac{-4}{(x+1)} + \dfrac{\frac{14}{3}}{(x+2)}}$

$x^2-3x+4 = A(x^2+3x+2) + B(x^2+x-2) + C(x^2-1)$
$\qquad\qquad = Ax^2 + 3Ax + 2A + Bx^2 + Bx - 2B + Cx^2 - C$
$\qquad\qquad = (A+B+C)x^2 + (3A+B)x + (2A-2B-C)$

$1 = A+B+C$
$-3 = 3A+B \qquad x=1, \quad x=-1, \quad x=-2$
$4 = 2A-2B-C$

if $x = 1$ solve for A
$(1-3+4) = A(1+3+2)$
$\dfrac{2}{6} = \dfrac{6}{6}A$
$\dfrac{1}{3} = A$

$x=-1$ solve for B
$1+3+4 = B(1-1-2)$
$8 = -2B$
$-4 = B$

$x=-2$ solve for C
$4+6+4 = C(4-1)$
$14 = 3C$
$\dfrac{14}{3} = C$

Figure 6

21

Next, you should ask yourself these general study questions to focus your attention on the general principles of the material just covered.

1. What are the different kinds of problems and how can they be recognized? Is it the *format* of the problem or the *directions* which will indicate the specific technique to be used?
2. How are these different problems related (or are they?)?
3. What other versions are possible for this type of problem? (How can this *problem* be restated?)
4. What are the different ways in which the directions can be worded and still mean the *same* thing?
5. What changes in the wording of the directions indicate *different* procedures?
6. Is there only one method to work this type of problem or are several techniques applicable? If several techniques are appropriate, how does one choose which to use (or does it matter?)?
7. What means (if any) are available to check your answer other than reworking the problem the same way?
8. How are the problems from this section/chapter *different* from the problems of previous sections/chapters?
9. How are the problems from this section/chapter *like* the problems of previous sections/chapters?

To fix firmly in your mind the new material you have learned, you should now with your books closed try to summarize in your own words the key features of the homework. That may mean focusing on new procedures for working problems or on how to recognize the different types of problems. Be sure to note any new vocabulary introduced. If questions come to mind, check in your notes or text. Record any unanswered questions.

Some people find it advantageous to read ahead in the text so that they are familiar with the new terminology that will be covered in the next class. You should feel free to ask your instructor what will be covered so that you can read the text prior to class.

Study Questions for Chapter 3

1. List several features of a good study environment.
2. What is one way to balance study time with breaks?
3. When reading *worked out* examples in the text, you should try to

   _____ .

4. What should you do *before* working problems for homework?
5. What should you do if you do not know how to begin a problem?
6. What should you do after you finish your homework?

Questions for the Student

1. Describe your study environment. Are there ways you can improve it?
2. What is the best time of day for you to study mathematics?
3. How do you organize your study time? Do you take breaks? Do you study some every day?
4. Have you made any study cards for yourself?
5. Do you read your text and notes before you begin the assignment?
6. Do you have someone with whom you study math?
7. Is there someone in class from whom you can get good notes if you are absent?
8. Do you look at your corrected homework and try to understand what you did wrong on the problems you worked incorrectly?

CHAPTER 4
Studying for Quizzes and Examinations

In most mathematics courses, you will be required to demonstrate your mathematical proficiency so that you can be assigned a grade. Usually this demonstration is a written test given in class, limited in time, and taken without the text. A large portion of your grade may be determined by your performance on these examinations, and, more than likely, you will not be given the opportunity to re-test if you do poorly. Of course, you have been preparing yourself for tests as you attend class, study your notes and text, and work assignments; however, you may not have learned the material thoroughly enough to perform well. Understanding the classwork or being able to do problems at home without help is no guarantee that you have learned the material thoroughly enough to earn a good grade on a test. Our goal in this chapter is to help you develop the necessary skills so you know the material well enough to perform effectively under pressure.

What is the difference between doing problems at home and taking a test?

> "But I did all the homework!"

When you do your homework, your time is more or less unlimited. You can take a break when you get stuck on a problem and do something else to clear your mind, then come back to the problem. Traditional tests require relatively quick responses with little time for reflective thought; you cannot take a break. When doing homework or classwork, you have notes or the text for reference; if you get stuck, you can look up a similar problem. You also may work with someone with whom you check answers and/or discuss solutions. Traditional tests do not allow references or a discussion with a friend—you are on your own.

Because most instructors tend to assign homework on material covered that day, you may recall enough of the class to be able to solve your homework problems. However, your test will cover many classes over a longer period of time. As you do homework, the text also provides you with subtle hints. Many textbooks group the problems according to how they are to be done: doing one problem in a section is knowing how to do them all. Knowing the section of the book your problem comes from gives you a large clue as to how to work the problem. Seeing the same type of problems grouped together also helps you determine their similarities and therefore cues you as to how they are to be solved. Most texts also provide answers to some problems and give worked-out examples which further help you. Sometimes the directions in the text are more explicit than are the directions on the test. Without any of these aids, test-taking becomes more difficult. As you do your homework, you may not realize that you depend upon these cues to help you solve the problems, and you therefore develop a false sense of security. You may believe you know the material, but when those crutches are not available and when you have limited time, you may not be sure how to work the problems.

If it looks familiar, don't you know it?

Most courses begin with a review of familiar material. Because the material is familiar, you may be tempted to "shut off" your mind and become impatient. It is crucial that you learn these basic skills *perfectly* before you go on to more complicated topics. Unless you can work the problems almost without having to think and without error, you do not *know* the basics well enough. Without this complete command of the fundamentals, your mind will not be free to concentrate on the new material. If you have any problems with the basics, see the instructor, a tutor, or a classmate; study, and restudy, but make certain that you master them! (If you genuinely believe that you cannot master the material, then you should discuss your placement with your instructor.)

Why is recognizing the problems important?

> "I got the problem right, I just did the wrong thing."

Many errors on math tests are a result of confusing one type of problem with another. For example, some students know that $\frac{ax}{x}$ can be reduced to "a" by cancelling like factors: $\frac{a\cancel{x}}{\cancel{x}} = a$. They then see the problem $\frac{a + x}{x}$, see the common terms of x, and reduce the problem: $\frac{a + \cancel{x}}{\cancel{x}} = a$, which is INCORRECT. These students are *familiar* with the material, but they do not *know* it: they think terms can be treated like factors.

Some books point out the kinds of problems you are likely to confuse. Many instructors help you to focus on these differences. It is important that you understand the distinctions between those problems which look to you to be the same, but which are, in fact, different. Examine the similarities and the differences, study them, and know how to distinguish one type from another. In making these distinctions you will find that you will gain an increased understanding of the topics you are studying.

Why does it matter when you study?

> "But I studied all night for this exam, and *still* flunked."

You must not expect to sit down the night before a test and learn all you need to know in one long study session. As we said in the introduction, learning mathematics is very much like mastering a sport or a language: you must practice consistently to become proficient. Therefore, you need to begin your study several days before a short quiz. Most successful students begin intensive study for a major examination a week and a half to two weeks before the test. Beginning the review for the test early allows you to study for several hours, then turn to other work. If you discover material you do not understand, you will have time to ask questions of the instructor or tutor. When you again study the math, you can see what you have remembered and what material still requires further study. By repeating this process several times a day for several days, you allow your mind to assimilate the material more efficiently and completely than it could if you tried to learn it in only one or two long study sessions. Give your mind a break! Start studying early!

What kind of activities are important when studying?

You must be sure your studying includes four basic activities:
1. Drill with note cards.
2. Practice working problems.
3. Review notes and text.
4. Reflect on similarities, differences, and possible variations on problems.

In any given study session, you should begin with one of the first three activities, then switch to another when you find your concentration slipping. In this way, you can vary your studying to keep your attention from wandering. As you end each session, you should allow some time to reflect on your earlier studying. As you become more proficient with the material, you should spend more time on reflection.

How do you write out study cards?

We have previously mentioned note cards; let us discuss the cards in greater detail. Using study cards is an efficient way to study for exams. Many people use 3" by 5" index cards, but some prefer 5" by 8" cards because more may be written on each card. In either case, you use them to summarize the key definitions, terms, and rules as well as to drill yourself to spot likely errors. Specifically, you will want a study card for:

1. *Every new definition or term used.* Be sure you have the word being defined spelled correctly at the top of the card. From your notes or the text carefully copy the definition onto the card with both an example of the definition and an example where the definition does not hold. You may want also to give a restatement of the definition in your own words.

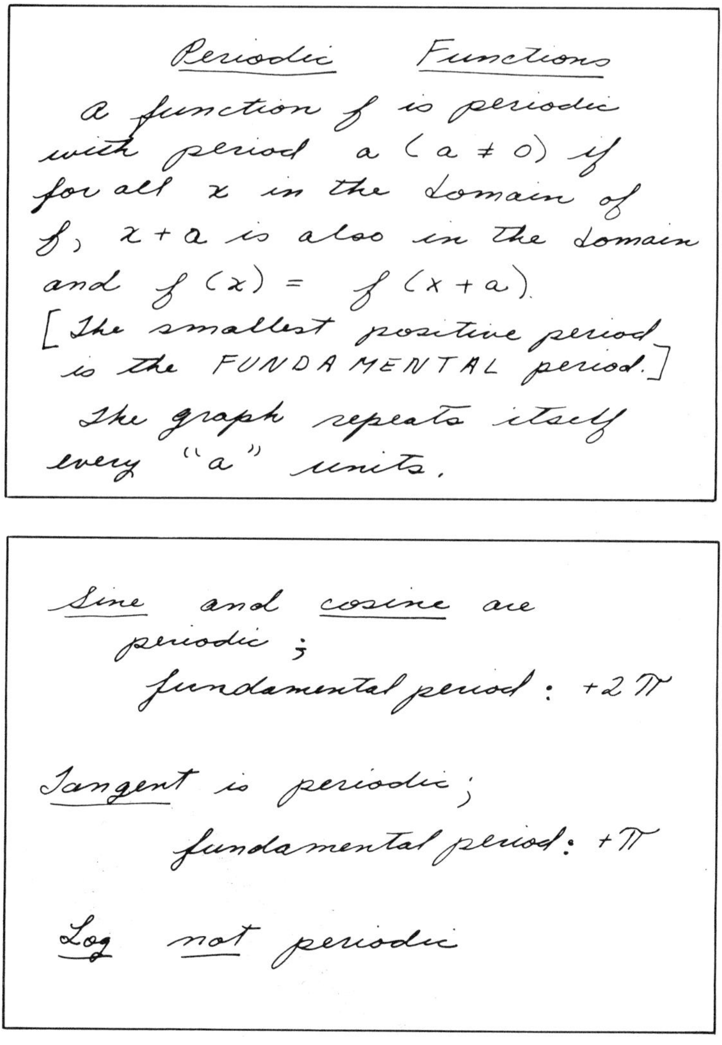

> **Periodic Functions**
>
> a function f is periodic with period a $(a \neq 0)$ if for all x in the domain of f, $x + a$ is also in the domain and $f(x) = f(x + a)$.
>
> [The smallest positive period is the FUNDAMENTAL period.]
>
> The graph repeats itself every "a" units.

> Sine and cosine are periodic;
>   fundamental period: $+2\pi$
>
> Tangent is periodic;
>   fundamental period: $+\pi$
>
> Log not periodic

Figure 7

27

2. *Each new procedure or algorithm.* Write each step of the procedure or algorithm. Be sure to re-read what you have written to make certain you have not omitted any steps. Copy an example of the procedure on the card. Many students prefer to use the back of the card for the worked-out example.

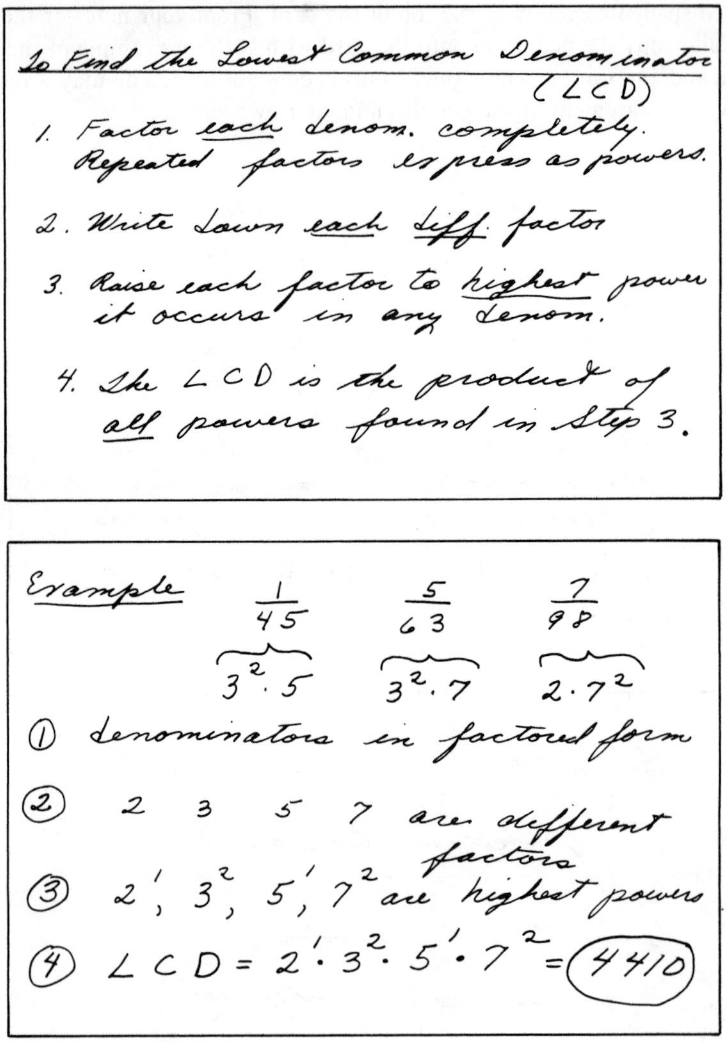

To Find the Lowest Common Denominator
(LCD)

1. Factor each denom. completely.
   Repeated factors express as powers.

2. Write down each diff. factor

3. Raise each factor to highest power
   it occurs in any denom.

4. The LCD is the product of
   all powers found in Step 3.

Example $\dfrac{1}{45}$ $\dfrac{5}{63}$ $\dfrac{7}{98}$

$\overbrace{3^2 \cdot 5}$ $\overbrace{3^2 \cdot 7}$ $\overbrace{2 \cdot 7^2}$

① denominators in factored form

② 2   3   5   7   are different
                          factors

③ $2^1, 3^2, 5^1, 7^2$ are highest powers

④ $LCD = 2^1 \cdot 3^2 \cdot 5^1 \cdot 7^2 = \boxed{4410}$

Figure 8

> "I got stuck on that same problem when I had it on last week's quiz."

3. *Common types of errors.* You should have a "warning card" for each error you find yourself repeatedly making. You can use corrected homework, quizzes, and exams to determine your common errors. You also should have a card for likely errors as pointed out in class or the text. Each card should give an example of the incorrect *and* correct procedures—*clearly* labelled!

To Reduce a Fraction to Lowest Terms

1. FACTOR the numerator and denominator completely.

2. DIVIDE numerator and denominator by all factors common to <u>both</u>

FACTORS c<u>an</u> be cancelled.

TERMS can<u>not</u> be cancelled.

Example: $\dfrac{x-3}{x^2-9} = \dfrac{\cancel{x-3}^{1}}{\cancel{(x-3)}(x+3)} = \dfrac{1}{x+3}$

Example: $\dfrac{x+y}{x}$ cannot be reduced: x is not a factor, it is a <u>term</u>

Example: $\dfrac{x+3}{x+6}$ neither x nor 3 is a factor; this CANNOT be reduced.

Figure 9

29

4. *Problems that give you great difficulty.* Most people find some types of problems especially difficult to master. If you find that you just cannot seem "to get the hang" of a particular type of problem, make out a card for it so you can drill yourself. (We assume that if you do not understand the problem, you have seen the instructor or a tutor for help.)

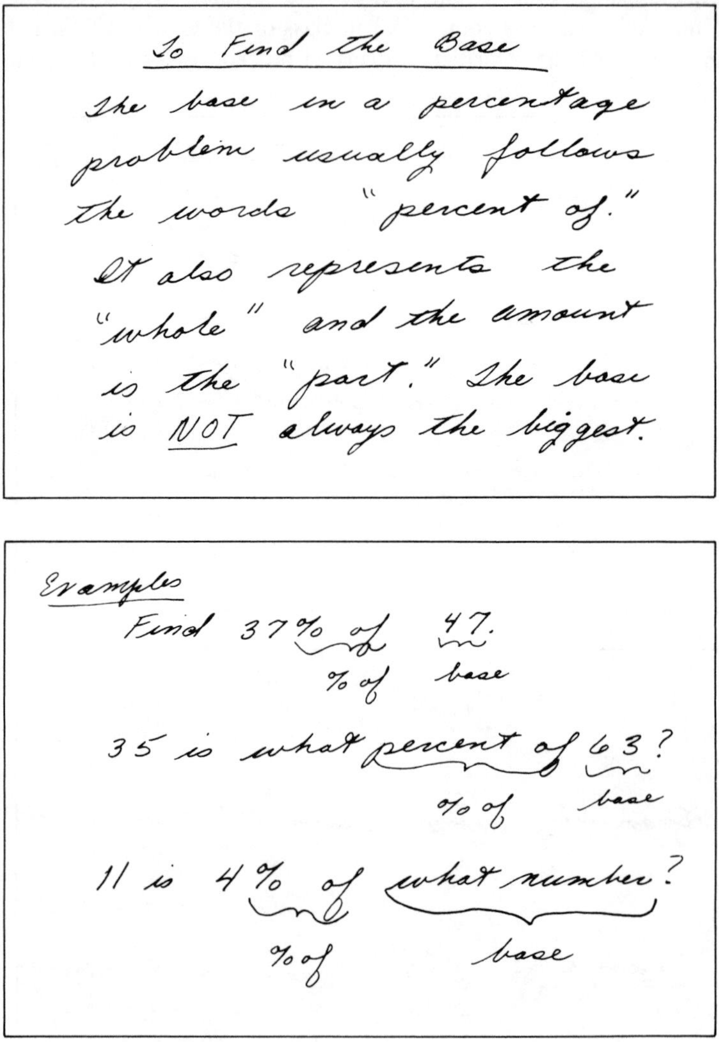

To Find the Base

The base in a percentage problem usually follows the words "percent of." It also represents the "whole" and the amount is the "part." The base is <u>NOT</u> always the biggest.

Examples

Find 37% of 47.
%of   base

35 is what percent of 63?
%of   base

11 is 4% of what number?
%of   base

Figure 10

30

5. *Quiz cards*. After you have studied the material thoroughly, you will want to test your knowledge. One way to do this is to have quiz cards made up to work. Select two or three problems from each exercise of the text to enter on cards. Copy the problem with directions onto the card. If answers are available, enter the answer on the back of the card. (You may also wish to note on the back of the card the page of the text containing the problem.) Other sources for problems are old quizzes and tests. Always be sure to copy the instructions for each problem.

How do the cards help you study?

Making the cards, of course, is just the first step in studying. Take the cards for a particular section or chapter and go through them one at a time, uncovering the top line or two so you read only the key word or words. Try to recite the rest of the information on the card without looking; then check to see if you were correct. Repeat this procedure for each card. As you go through the cards, you may wish to sort them into two stacks: one containing the cards you know and the other, the cards you must continue to study. When you feel you have mastered the cards from a single section or chapter, shuffle the deck of cards and go through them at odd times during the day when you have a few minutes free. Some of these times may be when you are waiting for the bus, the few minutes extra between classes, or when you want a break from studying for another course.

As you study the cards, ask yourself the following questions:

1. When would I use this formula?
2. How are these examples similar or different from other examples?
3. How can I distinguish the different types of problems?
4. To what type of problem does this definition, rule, or procedure apply?
5. What are the directions? Can I word them differently?
6. What different techniques can I use to solve this problem?
7. How do I know when to use the different techniques?
8. How can I check the answer?

For a comprehensive test, collect the cards for all of the sections and/or chapters covered and repeat the process. Be sure to keep reshuffling the deck as you improve. You should occasionally review those cards you have separated out to make sure you remember them and fully understand their relationship to the new material you are learning.

But you've done your homework; do you need to work more problems?

When you feel comfortable with the information on your cards or when you need to switch study activities, you should try to work problems from the appropriate exercises in the text. First, you should re-work any homework problems that were incorrect, then go on to other problems. If time permits, try problems that were not assigned. You should attempt the problems without looking at the examples or the discussion in the book. Refer to the text only when you feel that you really cannot determine how to work the problems.

Familiarize yourself with the directions; ask yourself if there are different ways in which the directions can be worded to get the same answer. Are there different directions appropriate for the problem that would result in different answers?

As you work problems, begin to focus on time. See how many problems you can work correctly in, say, twenty minutes. Because most tests have a time limit, you need to start working on speed as well as accuracy.

How can you check your work?

> "I would have passed if I didn't make so many dumb mistakes."

Nothing builds confidence more than knowing you are right, and the more confident you are, the more likely you are to persist. Most texts provide answers to many of the exercises, so you can check your answers (or answers to similar problems) as you work. You get hints as to what you may be doing wrong, and you gain confidence when you get them correct. Because you cannot have answers for problems on exams to give you hints as to how to do the problem correctly, you need to find other ways to build your confidence and help you determine the accuracy of your work. The importance of knowing how to check your work cannot be overestimated. Whenever you do a problem, always think about the ways you may establish the accuracy of the answer.

Solutions to equations may be checked by substituting the solution into the equation and determining whether both sides of the equation are equal.

You can verify answers for *word problems* by estimating the answer before you work the problem, by checking the units, and by re-reading the problem when you finish to see if your answer "makes sense."

You can check *graphing* by examining the characteristics of the equation to see if they are appropriate to the type of graph you have (e.g., a second degree equation usually cannot be graphed as a straight line; a positive coefficient for the highest power of the variable indicates that the graph increases and is positive for large positive values of the variable).

Factoring can be verified by multiplying together the factors.

Checking the *simplification of expressions* or of identities is more difficult. The only true check is to rework the problem; however, for a quick "guess" as to whether the problem is correctly worked, you may substitute a value for the variable in the original expression and in your answer; if they are NOT equal, then you made some error. If they ARE equal, however, you are not guaranteed that your work is correct. For instance, one student simplifies $\frac{x^2 + 4}{x + 2}$ to be $x + 2$. To check, he lets $x = 0$: $\frac{0^2 + 4}{0 + 2} = 2$ and $0 + 2 = 2$, which seems to indicate that the problem has been simplified correctly. That is not the case, however. Let $x = 2$, and we obtain $\frac{2^2 + 4}{2 + 2} = 2$, while $2 + 2 = 4$. Because $2 \neq 4$, we know that there is an error in the solution. Obviously, if you choose to use this method, you need to substitute in *at least* two different values for the variable. Usually you should choose numbers other than 0 or 1.

These are some of the major techniques used to check answers. *Because there are so many different techniques used to check problems, you should be alert for other methods presented by your instructor or in your text.*

Why is reviewing important?

Even though you may have all of the key information recorded on your note cards, you need to review your notes and the text. As you learn more about a subject, you become more sophisticated. When you go back through, you may discover information that you completely missed in your earlier studying. You are more ready to understand short cuts or tips than you were before. Reviewing can improve your grade by helping you become more efficient in recognizing problems and how to work them. As you study, be sure to focus on the relationship(s) among the various topics covered.

When studying figures that illustrate concepts, examine the figures with care; note every point. Figures are usually accompanied by verbal material. See if you can cover the figure with an index card and redraw the figure with only the verbal material as your guide. Cover the verbal material with an index card and rewrite the verbal material with only the figure as your guide.

What do you mean by "reflect"?

> "I can do okay on quizzes, but I mess up on major tests."

As we stated earlier in this chapter, one of the major difficulties students have is confusing one type of problem with another. On tests, in later courses, and in applications, you will need to determine the distinguishing characteristics of a given problem and the appropriate procedures for it. This is the most difficult task in test-taking. The way to prepare yourself for it is to anticipate the likely ways you may become confused and determine the clues that you must look for to avoid the confusion. The general questions to ask yourself as you review are the same questions that you used as you completed your homework. We repeat them here:

1. What are the different kinds of problems and how can they be recognized? Is it the format of the problem or the directions which will indicate the specific technique to be used?

2. How are these different problems related (or are they?)?

3. What are the different ways in which the directions can be worded and still mean the *same* thing?

4. What other versions are possible for this type of problem? (How can the *problem* be re-stated?)

5. What changes in the wording of the directions indicate *different* procedures?

6. Is there only one method to work this type of problem or are several techniques applicable? If several techniques are appropriate, how does one choose which to use (or does it matter?)?

7. What means (if any) are available to check your answer other than re-working the problem the same way?

8. How are the problems from this section/chapter *different* from the problems of previous sections/chapters?

9. How are the problems from this section/chapter *like* the problems of previous sections/chapters?

The study questions must necessarily be general to cover many different levels of mathematics. To indicate how you can prepare specific study questions for your course, we list below some study questions for arithmetic, for algebra, and for trigonometry.

Sample study questions for arithmetic

1. State the rounding rules.

2. State the correct order of operations.

3. When do you need to change mixed numbers to improper fractions?

4. State the procedures for finding an LCD (least common denominator).

5. For what kind of problem do you need to find an LCD?

6. In what type of problem do you reduce (cancel common factors)?

7. How do you recognize a complex fraction?

8. Give the steps needed to simplify a complex fraction.

9. State the steps for division of fractions.

10. How do you check your answer to a $\sqrt[]{}$ problem?

11. How do you tell if two fractions are equivalent (or two ratios form a proportion)?

12. In decimal multiplication, how do you determine how many decimal places there are in the answer?
13. How do you determine where to place the decimal in decimal division?
14. How do you multiply (or divide) a decimal by a power of 10?
15. How do you change a mixed number to a percent (without rounding)?
16. How do you change a percent to a fraction?
17. How do you identify the rate (percent), base, and percentage (amount) in a percent problem?
18. How can you use the units to help you set up a proportion?

Sample study questions for algebra

1. What is the difference between **TERMS** and **FACTORS?** (How do you recognize each?)
2. What are the rules for working with **TERMS?**
3. What are the rules for working with **FACTORS?**
4. Give the rules for exponents and when each is to be used.
5. Outline the different types of **EQUATIONS** solved in the course and the procedures appropriate to each.
6. Contrast the directions "Solve" and "Simplify." List the types of problems appropriate to each.
7. Outline how to work each type of problem listed in 5 above if you have not included it in a previous question.
8. What different results will you obtain when working with **EXPRESSIONS** versus **EQUATIONS?** How is the answer likely to be expressed in each case?

Sample study questions for trigonometry

1. What are the trigonometric functions for $\theta = 30°$, $45°$, $60°$, and $90°$?
2. How do you convert from radians to degrees and from degrees to radians? ($2\pi = $ _____ °)
3. Give the graphs for $\sin \theta$, $\cos \theta$, and $\tan \theta$ for $0 < \theta < 2\pi$ rad.
4. Which trigonometric functions are reciprocals of each other?
5. Which trigonometric functions have related graphs?
6. What happens to the values of the trigonometric functions as θ approaches $0°$, $90°$, $\frac{3\pi}{2}$, 2π?
7. Give three or more principal trigonometric identities.
8. What is the period of each trigonometric function?
9. What is the period of $y = a \cdot \sin(cx + d)$, of $y = a \cdot \cos(cx + d)$, of $y = a \cdot \tan(cx + d)$?

10. What is the amplitude of $y = a \cdot \sin(cx + d)$, of $y = a \cdot \cos(cx + d)$, of $y = a \cdot \tan(cx + d)$?
11. What is the translation of $y = a \cdot \sin(cx + d)$, of $y = a \cdot \cos(cx + d)$, of $y = a \cdot \tan(cx + d)$?
12. For what values of θ (if any) is each trigonometric function undefined?
13. What does arc sin θ or $\sin^{-1}\theta$ mean?
14. Graphically how are the trigonometric functions related to their inverses?

How do you know when you are prepared?

> "But I get the problems right when I do them at home."

> "I could work it if they gave me more time."

There is no way to know for a certainty that you are prepared for a test; however, there are ways to check your preparedness: take a practice test as if it were the real test and see how you do. If you are satisfied with your performance, then you can relax knowing you have a reasonable command of the material. If, however, there are problems you cannot work, or if you do not come near finishing the test in the allotted time, then you know you must resume your study to improve in your weakest areas. All too frequently, good students who have mastered the material fail to acknowledge the time factor of a test. They have never thought about the time they spend working problems and too often find themselves with problems untouched when the time is up. Needless to say, this situation is extremely frustrating and will inhibit the concentration of the students on the next test unless they squarely confront the time limitations as they study.

In some courses old tests and quizzes are available for you to use as practice tests. The quiz cards are another source of problems on which to test yourself. If you have a "study buddy," you can make up tests for each other. Whenever you work a practice test be sure you simulate as nearly as possible the actual test limitations. Go to a quiet corner of the library, disconnect the phone, do whatever you need to secure an uninterrupted time. Set an alarm or timer for the appropriate period and then set to work—without notes, books or any item except those which will be allowed at the actual test. If you take the test with a "study buddy," do not exchange **ANY** information or looks during the practice test. You can compare answers after the time is up. When your time is up, **STOP.** Check your answers if you have the answers available. If not, evaluate how you think you did. (If there is time, have a classmate, tutor, or instructor check your work.) Did you work efficiently? Were you able to finish the test in the allotted time? If not, do you think the problems you did are correct? If you cannot finish a test, then you should at least be reasonably certain that the problems that you *did* do are correct. Reflect on the way you approached the test. If you have doubts about your efficiency, reread Chapter 5 to determine what changes you need to make in your test-taking.

If you were not satisfied with your performance, then study the material you missed or revise your test-taking strategy. When you again think you are ready, try another practice test.

Non-traditional exams

Up to now we have only discussed traditional exams: in-class, closed book, written tests which are limited in time. Some of you may need to prepare for other types of tests: open book, take home, unlimited time, and mastery. Non-traditional tests also require more intensive study than that needed to do homework. Do not believe that open book or take-home exams require less effort than traditional tests: they can be more difficult. The instructor usually assumes that you know where to find the formulas and procedures and may give you problems requiring more sophisticated math skills.

In-class, open-book, or open-note tests

The most important thing to keep in mind when preparing for an open-book or open-note exam is that the less time you spend leafing through book and notes, the more time you can spend answering the exam questions. Thus, you need to be thoroughly familiar with your book and notes: know exactly where to find the information you need. If you tab your book or notes at pages containing key formulas and procedures, you can find them more quickly. If there are many important topics, then tab only the most important pages and familiarize yourself with the book index so that you can turn to the needed topic quickly.

In some instances your instructor will permit you to bring only one file card or one sheet of paper into the test. In this case, you should carefully outline for yourself the material to be covered on the test and determine that information which is most crucial that you know. Once you have made these decisions, then try to get that material recorded on the card or sheet. Be sure you have it well-organized so that you can find the needed formula or fact easily.

You should study for this exam as though you were studying for a traditional test. You need to understand concepts and have practiced working problems. You may not need to memorize the exact formulas or procedures, but you need to have them readily available.

Take-home exams

The take-home exam usually has more difficult problems than in-class tests. Because the student has plenty of time and resources available, more challenging problems are appropriate. The student who has been studying consistently again has an advantage over the procrastinator. Do not postpone your studying until you receive the test; you may not have enough time to both learn the material and complete the test.

In-class tests with unlimited time

Studying for this type of exam is much the same as studying for traditional exams. Time is less important than is your success in disciplining yourself to use the time available.

Re-take exams and mastery testing

Frequently when students know that re-tests are offered, they do not properly prepare for the initial test. They then must try to master the material in the (usually short) time between the test and the re-test. This is not the most constructive way to study. Instead, study for the first exam as though you will not have a second opportunity. If it is to your advantage to take the re-test, then your review should require less time. You obviously should review thoroughly the topics which gave you difficulty on the first test.

Study Questions for Chapter 4

1. List some advantages you have while working problems at home that you do not have when taking a test.
2. Many errors on tests are a result of _____ .
3. When should you start studying for a test?
4. What are the advantages of using study cards?
5. When using study cards, what are some questions you should ask yourself?
6. What are some ways you can check the accuracy of your answers on an exam?
7. Open-book exams are always easier than closed-book exams. (True or false.)
8. In open-book or open-note tests, it is important to _____ .
9. How do you choose information to record on a study card?

Questions for the Student

1. How do you drill yourself to recognize problems?
2. Have you made study cards for yourself?

3. What is the most common error you make? How are you trying to overcome this error?

4. Do you ever work more problems than those assigned?

5. Do you know how you can check your work?

6. What additional methods for checking has your instructor given you?

7. What kinds of tests do you take in your current math course? How do you study for them?

8. When preparing for an exam, do you work problems with the book or your notes open?

9. Do you ever time yourself as you work problems to prepare for a test?

10. Is there a difference between the study habits you needed for high school and those you need in your current course? If so, what do you need to do differently?

11. How do you judge whether or not you are adequately prepared for a test?

CHAPTER 5
Taking Quizzes and Examinations

Are you ready?

It is hard to determine when an individual is ready to take a test. Knowing the material is, of course, a major factor in determining how well you can perform, but it is not the only factor. Knowing the material sets the upper limits; other factors such as lack of sleep, hunger, or anxiety can impair your performance.

If you followed the steps outlined in the previous chapters, you should know the material well enough to take the test. However, if you find that you still tend to make too many careless errors, tend to draw blanks, or if the person you helped in your class did better than you on your last test, then you may have a problem taking tests. This chapter will help you improve your test-taking skills so you can more clearly demonstrate your full knowledge.

What should you do before the test?

Get a good night's sleep before the test. "Pulling all nighters" (studying all night) for math exams seldom works. If you do not get enough sleep, you increase the probability of making careless errors or "drawings blanks." Some students who stay up all night studying end up falling asleep during the test. Others try to stay awake by taking "NoDOZ" or other pills; the drugs make the students so nervous that they are unable to concentrate on the test and, consequently, are unable to demonstrate that which they *do* know. Recent studies indicate that sleep performs an important function in the learning process: it gives your brain the chance to consolidate or assimilate that which you have just learned.

You should also eat properly: it is hard to concentrate when you are hungry. To avoid another deterrent to concentration, make sure you go to the bathroom before the exam!

Do not schedule an appointment immediately after the exam. Your concern about meeting it will increase your anxiety in the test. Furthermore, if your instructor decides to give you an extra few minutes, your worry about missing the appointment will decrease your concentration on the test. If you have a class right after the exam, try to inform that instructor in advance that you may be late. It is very important that you have no other concerns but the exam during the exam.

Do *not* study right up to test time. Realizing that you forgot to study a topic or that you cannot do a problem five minutes before the test may cause you to panic. Then you will do poorly on the whole test, rather than just missing the one or two test questions on the material you forgot to study. Take at least three hours prior to the test to relax and engage in activities other than studying for the test. Studies have shown this relaxation period is an important part of the learning process.

It is advisable, however, just before the exam, to refresh your memory on *one* or *two* particular procedures or formulas you wish to remember. You should also check to see that you have the necessary materials for the test: pencils, eraser, pens, ruler, pencil sharpener, charged batteries in your calculator (if it is permitted), or other items.

What is the best procedure in the test?

As soon as you are told to begin, on the back of the test jot down any formulas or key ideas that you do not wish to forget. You may want to write a warning such as: $a^x \neq x^a$ to remind yourself that these expressions require different formulas for differentiation (for calculus), the quadratic formula (for algebra), or important trigonometric identities (for trigonometry). You can then relax knowing you have them recorded. When you have finished making your notes, write your name on the paper and begin the exam.

First, *read the test directions carefully.* This is especially crucial if the exam is computer graded. Starting in the wrong column may make your 100% into a 0%. If the directions say "circle your answer" or "reduce all fractions," then do just that! Read the directions for all problems carefully. When you finish the problem, re-read the directions to make sure you gave the answer requested in proper form.

If you have any problem understanding the directions, do not hesitate to ask the instructor or proctor for clarification. It is better to ask than to guess at what the problem requires and get it wrong. The worst that can happen is that the proctor will tell you that he/she cannot answer your question.

Be sure to pay attention to any announcements written on the blackboard or made out loud by the proctor. That is the one time you should interrupt your concentration on the test!

Although many instructors arrange the test problems in order of difficulty (easy problems first), not all teachers do that. Besides, what may be easy for someone else may not necessarily be easy for you. There is no reason why you should do problems in the order they are given on the test. If the more difficult problems are at the beginning of the exam, and you spend a lot of time on them, you may not have the time to work on the easier problems that come later. As a result, you end up with a lower grade than if you had started with the easier problems.

In order to use the time efficiently, we recommend that you work test problems in the following order:

1. Do first the problems that are easy and quickly solved and that you know how to do.
2. Do the problems that you know how to do which may take a little longer.
3. Attempt the problems for which you have some idea about solving.
4. Allot the remaining time between checking your completed work and attempting the rest of the problems.

Do not spend an excessive amount of time on any one problem unless you are completely satisfied with your work on all the other problems.

> "If I could just get started on the problem then I'd be okay."

You do not need to know how to finish a problem to start it! If you are not sure how to proceed, write down all that you know about the problem: any relevant formulas, facts, or procedures. Think about problems you have done that look similar to this one. Frequently just doing that much will enable you to continue the problem. If you still do not know what to do next, circle the problem number and go on to another one. Frequently, if you have time to come back, your notes may jog your memory, and you will be able to work the problem. If not, then you know you have not wasted time on a problem that you do not understand.

How should you go about working a problem?

Check each calculation as you do it. You may be able to check more quickly using the inverse operation (check subtraction with addition or division by multiplication). Make sure you did not make a copying error as you take each step.

After you finish each problem, re-read the question to see if your answer makes sense. Do the units match? Have you answered the question? Re-read the directions. Be sure your answer is labelled correctly. Can the person scoring your test find your answer? If you wrote down a warning of an error you tend to make such as $(-2)^2 \neq -2^2$, you should refer to your note each time you work a problem in which you might make that error.

Use the checking techniques discussed in the previous chapter. Try not to do the problem over in the same way. If your answer does not agree with your check, see if you can quickly spot any errors made in the logic you used or steps you took. Perhaps you made a numerical error, copied a step over wrong, or used the wrong formula. If you still cannot spot what you did wrong, circle the number of the problem so you can remember to come back to it. If time permits, rework the problem *from the beginning.*

How much time should you spend on a problem?

Speed and accuracy are both important for traditional exams. Do not spend an excessive amount of time on any one problem. To determine approximately how much time to spend on each problem, divide the allotted time by the number of problems to be worked. If you have 75 minutes to do 27 problems, you should spend, on the average, $75 \div 27 = 2.8$ minutes per problem. This estimate, however, assumes each problem counts equally. If some problems are worth more than others, you can safely spend more time on those with greater point value and less on the others. In all cases, do the problems in the order outlined in the previous section.

If your time is up, you must turn in the exam; however, unless the time limit is really restrictive, many students will turn in their tests before the time is up. Because of anxiety, frustration, or maybe exhaustion, many may be tempted to turn in the exam too soon. There is no reason why you should not use the entire time you are given. When you feel that you have done all you can (including checking your answers) and you still have time left, turn the test over then relax for 2–5 minutes: one way is to think of something else and close your eyes and rest. At the end of your rest, go back and check your work.

What is test anxiety?

It is not uncommon for people to be anxious when they have something at stake on their performance. The difference between a rookie and a veteran is the veteran's ability to perform well consistently under pressure. The rookie baseball player may have a better batting average than the veteran, but when the pressure is on and a hit is needed, the veteran player will be called on to do the job.

While there is obviously a difference between the pressures faced by a baseball player in a game and the pressure you face in a mathematics test, the resulting anxiety may be as debilitating to your performance as it is to his: it is a matter of ability, confidence, and experience.

The steps outlined in the previous chapters are designed to help you develop your mathematical skills. Chapter 4 specifically helps you develop your ability to perform under pressure. If you did well on the quiz you gave yourself, you should feel more confident. If you quizzed yourself under conditions similar to the test you are to take, you should have enough experience to reduce your anxiety. The math test may still make you somewhat anxious, but that is okay. A little anxiety should not hurt you and may keep you alert.

> "No matter how well I know the stuff, I just can't do well on a math exam."

Here we are concerned with anxiety at a level which interferes with your performance: when you are so anxious you cannot think straight or you forget the "easy material."

We assume you have had enough rest, have eaten properly, and have relaxed a few hours before the test. Now you have the exam. The important thing is concentration—to be able to attend to the exam and the problems, and nothing else.

First, do not be distracted by other students. In particular, do not be concerned if other students leave earlier than you. Often students panic when they see others leaving the test early. You should be aware, however, that the students who leave first are usually not the ones who get the best grades; quite often, they are the students who have given up. Again, take all the time you are given. Try not to be the first to leave the test.

> "I panic on tests!"

What do you do when panic sets in? The key here is to rid yourself of internal distractions and to concentrate on the problems. People who panic on tests are too busy telling themselves how poorly they are doing, or how much they do not know, rather than concentrating on the problems. For example, let us assume Joe is taking an exam and doing well. Suddenly he comes to a problem he has never seen before. A typical conversation with himself goes as follows:

"Okay, I finished that problem, now let me go on to the next one."
"Let's see, problem number 15, find the volume under a hyperbolic paraboloid if . . ."
"A WHAT?"
"I don't know what a hyperbolic paraboloid is."
"What am I going to do?"
"I can't do this problem!"
"There goes my B."
"I can't do any of this stuff."
"I'm running out of time."
"I can't think!"

And on he went. He spent more time telling himself he could not do the problem than he spent concentrating on the problem and trying to reason out exactly what it was he had to do or what a hyperbolic paraboloid was. The danger here is that Joe could become so shaken by problem 15 that he may continue to make these negative self-statements throughout the test, destroy his concentration, and reduce his performance.

The initial panic at not being able to do a problem can start a vicious cycle which may be hard to stop. However, you can prevent the cycle from getting started. Once Joe saw that the problem was difficult to understand, he should have said to himself, "I'll come back to it later," and then gone on to the next problem.

If panic has already started and you find yourself so involved in negative self-statements that you cannot concentrate, break that cycle by yelling to yourself, **"STOP!"** Now, pause a few moments, try to relax, and clear your mind. You want to restore your concentration on the test. Take a few slow deep breaths and then look for a problem you are reasonably certain you can do. Once you complete that problem, continue to work on other problems on the test, following the order listed on page 42.

> "You didn't show us how to do *that* problem in class."

Do not worry if you see a problem you have never seen before. Many teachers give a few problems that require skills a bit more sophisticated than those possessed by many students. You may be able to solve it when you come back to it later (if you have time).

Can you practice the text anxiety procedure?

If you do panic during exams and have never tried the procedure above, it is a good idea to practice it before you take the test. You could do this by talking out loud while you do the practice tests and listening to yourself. If you tend to make statements like "This is too hard," or "I'll never do this," list those statements. Then create a list of responses to them which will help you get back to concentrating on the problem. Then practice those responses. For example, if you tend to say, "I'll never do this problem," your response may be, "I've done a problem like this before," "Now how did I do that other problem?" If you say, "I can't do math," you may respond with "I'm only having difficulty with this problem; now what part of the problem am I having difficulty with?" There is nothing difficult in this procedure for test anxiety. It is just a formal way of talking yourself into focusing on the problem rather than on yourself. You should practice the technique of yelling "Stop" to yourself as well.

What do you do if you freeze at the beginning of the test?

At the beginning of the test, you should be concentrating on the procedures we discussed in the previous sections: jotting down things to remember, reading directions, etc. If you find yourself panicking, then pause, clear your mind, relax, take a few slow, deep breaths and then start the problems following the order recommended on page 42.

Many times what students believe is poor performance due to test anxiety is really caused by a lack of preparation. However, there are students who completely blank out on tests (because of the pressure involved). If you are one and if the test anxiety procedures discussed above are not working for you, then you may want to consult a counselor for additional help.

Show-your-work tests

If you are to show your work, be sure it is organized so that anyone reading the test can follow your steps and can find your answers. This is crucial if partial credit is given. If you have the time, do some work on every problem. It does not hurt to guess at how to approach a problem or follow a hunch about an equation or an answer. However, use these approaches to problems you are unsure of after you have worked carefully on those problems you know how to do.

If a problem has several parts and you are unable to answer the first part, do not skip the rest of the problem! Make a reasonable guess or approximation for the first answer and proceed with the other part. For instance, you could be given the following problem (from arithmetic):

   a. Find the lowest common denominator for $\frac{3}{16}, \frac{5}{24}$, and $\frac{7}{18}$.

   b. Add the fractions.

You may not remember how to find the *lowest* common denominator (144) but you figure out that 288 will work as a denominator. Use it to work the second part of the problem.

Or you could be given a problem like:

   Compute the volume of an empty oil drum 6 feet high and 3 feet in diameter. If oil is pumped in at 12 ft³/min., how long will it take to fill?

Here you may not remember how to calculate the volume of a drum, or you may not recognize it as a type of cylinder. The volume, however, is necessary to answer the second question. You should then approximate the volume and use that figure to answer the second question. Be sure to state that the figure you used for the volume is your approximation.

Multiple-choice tests

On multiple-choice tests, always examine the form in which the answer is given so that your calculations will be appropriate. You will save time by working toward the form given rather than to another form and then having to convert. For instance, you would want to know whether the answer is given as a decimal or as a fraction. In algebra you would want to know whether an answer is factored or multiplied out. Be sure to check units and the form of algebraic expressions, fractions, and trigonometric identities. Make sure that you look at each possible alternative and see that your answer matches one of them exactly. However, do not assume that because your answer does match an answer given that your answer is correct! Generally the incorrect alternatives are created by using the most frequently given wrong answers. Therefore, you should still check your answer.

It may be quicker to use the answers to help you work the problem. First try to eliminate the obvious wrong answers and work backwards from the answers which are left. This technique can also help you on problems you do not know how to solve. Again, try to eliminate any obvious wrong answers. If you can eliminate several, then it may be to your advantage to take a guess as to which of the remaining answers is correct.

In-class test with unlimited time

While you may be given unlimited time for the test, in reality there is a limit to the time you can concentrate on the test. You should take the test as if your time is limited. When you finish taking the test using the procedures outlined previously in this chapter, take a 2–5 minute rest period, then check your work.

In-class open-book or open-note tests

You should have tabbed your book or notes and have a quick way of locating information. Otherwise, follow the steps given in the previous sections of this chapter.

Take-home tests

You should be careful not to procrastinate on take-home tests. In general they are much more difficult than in-class tests. You should begin the test as soon as possible and work through it in the order recommended in the previous sections. The sooner you start, the better. If you can complete the test a few days before it is due, put it away for a day and come back to it to check each problem carefully.

Why should you keep your tests?

You should keep your copies of all quizzes and tests that you take so that you may study them. You should note any errors that you made and rework (or learn how to work) any problems that you missed. You should think about *why* you missed problems or parts of problems. Did you forget a principle? Make a mistake in calculating? Confuse two types of problems? Was there an unexpected type of problem on the test? If so, can you anticipate similar problems on future tests? After evaluating your performance on the test, make out a study card for each problem missed (see pages 29 and 30).

If you have a question about the grade on a test, check with the instructor as soon as possible. You should keep a record of your grades and know how your final grade will be computed. You will then be able to check to see if any errors were made in assigning your course grade.

Study Questions for Chapter 5

1. What should you do the night before an exam?
2. When should you stop studying for a test?
3. After finishing a problem on an exam you should _____ .
4. On an exam, what order should you follow in working problems?
5. On the average, how much time should you spend per problem on an exam with a time limit?
6. If you have time remaining when you finish an exam, what should you do?
7. When should you start to work on a take-home exam?
8. How can you take advantage of the answers given on a multiple-choice question?
9. How can you help yourself remember formulas in a test?

Questions for the Student

1. When do you stop studying for your tests?
2. When you finish an exam, are you in a hurry to turn it in?
3. Do you jot notes to yourself on the test before you start working it?
4. If you find yourself getting nervous on a test, how do you handle it?
5. Do you keep your tests when they are returned to you?

CHAPTER 6
Self-Paced Study

Up to now we have discussed how to study and learn mathematics in a traditional course in which the instructor sets the pace and the class covers the same material at the same time. Here we will discuss study skills for non-traditional courses. The student may have more control over the pace. The primary source of information may be self-study guides, videotapes, film strips, slides, tape recordings, computers, or tutors which may be available in a math lab or learning center rather than a traditional classroom. Alternatively, a student could be enrolled in a correspondence course where the entire course is studied independently.

How do you pace yourself?

For most self-paced courses there is some time limit set for successful completion—it may be longer than the usual semester, but is usually less than a year. Because the allowable time period can vary so greatly, it is important for each student enrolled in such a course to have a very clear understanding of the time limitations of his or her particular course and how different completion dates might affect the grade for the course. You need to have a clear understanding of what material you are expected to have mastered within that period. You and your tutor or instructor can then set up a realistic timetable for covering the required chapters or units. Once the timetable is established, try to stick to it and study math some every day. If you find you are falling behind, be sure to discuss your progress with the tutor or instructor. Perhaps he/she can suggest different materials for you to use.

How do you get help?

Studying independently requires a lot of self-discipline. You must discipline yourself to keep working rather than put off studying. In some situations you must assume the responsibility to seek out a tutor who can help you with the material. Just remember that the tutors and instructors are there to help you learn mathematics. The self-paced structure was chosen in hopes that you would be more willing to ask questions.

If you think that the course materials are inappropriate for you (too easy or too hard) discuss your feelings with the instructor or tutor. If the materials are too easy, perhaps you can test out of a unit or two. If the material seems difficult for you, the tutor or instructor may have more appropriate supplemental material prepared that you may use.

Learning primarily from written materials

If you are working primarily from a text and/or worksheets, you probably should outline for yourself the major topics covered and make study cards for any areas in which you have difficulty. You should follow the suggestions of Chapter 3 on studying notes and text, and then working assignments.

Learning primarily from other materials

If you are learning primarily from videotapes, cassette tapes, or film strips, you may wish to go through each twice. The first time listen and read to understand, the second time to take notes on new and/or difficult material. You may wish to read Chapter 2 to improve your note-taking skills.

If you are learning primarily by using a computer, you need to make notes for yourself at the end of each session to summarize the information that you just learned.

Summary

No matter how you are learning math you should include outlining and making study cards. Writing is one way to get information into memory: the process of writing and outlining helps to organize thoughts. In addition, the outline is easier to remember.

Even though you may be learning mathematics independent of a formal lecture class, you still will need to learn the meanings of new terms that you encounter. You should modify the procedures suggested in Chapters 1 through 5 to your specific situation. In particular you should follow the suggestions in Chapter 3 for working assignments. As you work through assignments you may find problems that you do not know how to work. First make a quick review of the related work to make sure you did not skip an example that is similar. If you find no similar example, then contact your tutor or instructor for help. Do not be intimidated if the tutor or instructor can solve the problem quickly. They have had to work many, many similar problems and, therefore, are much more practiced than you.

To help you remember what you have learned earlier in the course, be sure to review regularly the earlier chapters/units that you have covered.

Study Questions for Chapter 6

1. How can you set goals in a self-paced class?
2. What is an advantage that videotapes, cassette tapes, and film strips have that a lecture does not have?

Questions for the Student

1. If you are in a self-paced course, how long may you take to complete the course?
2. When you have questions, where can you go to get them answered?